Analysis Report on
Fire Fighter Fatalities

Prepared by

Fire Analysis Division
National Fire Protection Association
Batterymarch Park
Quincy, MA 02269

Prepared for

Federal Emergency Management Agency
U.S. Fire Administration
Contract No. EMW-86-C-2278

August 1987

ACKNOWLEDGEMENTS

This study was funded by the U.S. Fire Administration of the Federal Emergency Management Agency (FEMA). It would not have been possible without the cooperation and assistance of the U.S. fire service, the United States Fire Administration, the Public Safety Officers' Benefits Program of the Department of Justice, the Bureau of Land Management of the Department of the Interior, the Bureau of Indian Affairs, the U.S. Department of Energy and the U.S. military, all of whom contributed data to this study.

Special analyses included in this report were authored by Dr. John Hall, Director, Fire Analysis Division and Mark Picher, Fire Data Technical Coder. Medical review of the heart attack study was provided by Dr. Allen J. Criss and Dr. Stephen R. Levisohn.

The assistance of other NFPA staff members is also acknowledged: A. Elwood Willey, Assistant Vice President, Research and Fire Information Services; Martin Henry, Director, Public Protection Division; Michael Karter, Ken Taylor and Paul LeBlanc, Fire Analysis Division; and Robert Barr and Carl Peterson, Public Protection Division.

The authors also wish to acknowledge the assistance of USFA staff, particularly John Ottoson. Project Officer, for their suggestions and support.

Special thanks go to Nancy Schwartz for typing the several drafts of this report.

Rita F. Fahy
Arthur E. Washburn
Project Staff

TABLE OF CONTENTS

LIST OF FIGURES

LIST OF TABLES

For more than a decade, the National Fire Protection Association (NFPA) has developed the most complete records on U.S. fire fighter fatalities - both in breadth of coverage and depth of detail - of any organization. This data base has been used to support the fatality studies produced each year by NFPA since 1974.

Over the past six years, NFPA also has worked with FEMA's U.S. Fire Administration (USFA) to provide, in a timely manner, lists of fire fighter fatalities and their next of kin to support the National Fire Academy's annual Fire Fighter Memorial Service. Under the present contract, NFPA has provided the USFA with lists, both hand lettered and typed, of 1986 fire fighter fatalities and with a list of names and addresses of next of kin and of fire department chiefs for use in the Memorial Service in October 1987.

In August, a briefing on the 1986 experience and three special analyses was presented by NFPA staff to USFA staff in Emmitsburg, MD. Through the briefing and analysis, this contact continued the trend toward more extensive analysis of patterns and trends in specific parts of the fire fighter fatality problem. With a full decade of experience now coded, NFPA is able to provide increasingly detailed and focused examinations of the specific parts of the problem addressable by particular strategies.

The deliverables under this contract are (a) this analysis report, (b) the incident and casualty data on tape in NFIRS Version 4.0 format, which is being delivered separately, (c) the various lists described above, and (d) the briefing provided in August.

I. INTRODUCTION

The purpose of this study is to analyze the circumstances surrounding fire fighter fatalities in the United States in 1986 in an attempt to identify potential means for reducing the number of deaths that occur each year. In addition to the 1986 findings, this study will also include special analyses of particular recurring scenarios, using NFPA's data base of fire fighter fatalities from 1977 through 1986.

A. Who Is a Fire Fighter?

For the purpose of this study, the term "fire fighter" covers all members of organized fire departments, whether career, volunteer or mixed; full-time public service officers acting as fire fighters; state and federal government fire service personnel; temporary fire suppression personnel operating under official auspices of one of the above; and privately employed fire fighters including trained members of industrial or institutional fire brigades, whether full- or part-time.

Under this definition, the study includes not just municipal fire fighters, but also seasonal and full-time employees of the U.S. Forest Service; prison inmates serving on state forest service crews; fire fighters for the Bureau of Land Management, the Bureau of Indian Affairs, and the U.S. Department of Energy; military personnel performing assigned fire suppression activities; civilian fire fighters working at military installations; and members of industrial fire brigades.

B. What Constitutes an On-Duty Fatality?

The term "on-duty" refers to being at the scene of an alarm, whether a fire or non-fire incident; being enroute while responding to or returning from an alarm; performing other assigned duties such as training, maintenance, public education, inspection, investigations, court testimony and fund raising; performing non-fire duties on official assignment; and being on call, under orders or on stand-by duty other than at home or at the individual's place of business.

On-duty fatalities include any injury sustained while on duty that proves fatal; any illness that was incurred as a result of actions while on duty that proves fatal; and fatal mishaps involving occupational hazards that occur while on duty. The types of injuries included in the first category are mainly those that occur on the fire ground, in training or in accidents while responding to or returning from alarms. The most common example of fatal illnesses incurred on duty is fatal heart attacks while on duty. Another example is a fire fighter who contracted hepatitis when a victim being transported by ambulance pulled out his intravenous needle and stuck the fire fighter. A few examples of fatal occupational mishaps include fire fighters who died of asphyxiation while working on fire apparatus in closed garages; a fire fighter who fell through a slide pole hole; a fire fighter electrocuted while raising a banner for a town event; a volunteer fire fighter who was fatally injured when he fell down a flight of stairs in his home while responding to an alarm; and a fire inspector who fell through a skylight.

Also included in the study are fire fighters who were murdered while on duty. These include fire fighters shot by snipers while on the fire ground, fire fighters shot in the station by off-duty or former fire fighters, and an arson investigator shot while interviewing a subject.

Fatal injuries and illnesses are included even in cases where death is considerably delayed. When the onset of the condition and death occur in different years, the incident is counted on the basis of the former. For example, a Michigan fire fighter died in 1986 of a brain injury received in 1979 when he was struck by a brass hose coupling, resulting in recurring seizures. Because his death was the direct result of his injury, and the injury occurred in 1979, he is counted as a 1979 fatality.

The NFPA recognizes that these definitions should include chronic illnesses (such as cancer) that prove fatal and that arise from occupational factors. In practice, there is as yet no mechanism for identifying fatalities that are due to illnesses that develop over long periods of time and that thereby present an ambiguous picture on the issue of occupational versus other factors as causes. This is recognized as a gap that cannot now be filled because of the limitations of the state of the art in tracking and analysis.

C. Sources of Initial Notification

As an integral part of its ongoing program to collect and analyze fire data, NFPA solicits information on fire fighter fatalities from the U.S. fire service and a wide range of other sources. These include the U.S. Fire Administration and the Public Safety Officers' Benefits Program (PSOB). Both are organizations with whom NFPA has maintained long-standing cooperative efforts in collecting and analyzing fire fighter fatality data. Other contacts include federal agencies such as the U.S. Forest Service of the Department of Agriculture, the Bureau of Land Management of the Department of Interior, the Bureau of Indian Affairs, the U.S. military and the Department of Energy.

We also receive notification from fire service organizations such as the International Association of Fire Fighters, state fire associations, state training organizations, state and local fire marshals, and fire service, publications. A network developed over the years of individuals interested in the area of fire fighter fatalities also assists in identifying incidents, especially those that occur outside of large urban areas or that involve non-fire-incident- related fatalities. Among these individuals are fire fighters, photographers, fire buffs, and members of the insurance, industry.

Notification of fatal incidents also comes from NFPA members and staff and through the use of a newspaper clipping service that reads all daily and weekly newspapers in the country.

D. Procedure for Including a Fatality in the Study

After initial notification of a fatal incident is received, contact with the local fire department is made by telephone to verify the incident, its location and the fire department involved. Data collection forms for the fatality and the fire, if it was a fire incident, are sent to the responsible local official identified during the telephone follow-up. After the forms are returned to NFPA, a final decision is made to include or exclude the fatality, based on the inclusion criteria described previously. In order to make a final determination, additional information is sometimes sought, either by contacting the fire department directly to clarify some of the details or by obtaining data elsewhere, such as medical documentation frequently available from PSOB.

Some of the material that might be received to document an incident includes casualty forms, both NFPA fire fighter fatality study reporting forms and NFIRS-type forms; NFPA's Fire Incident Data Organization major-fire report form or the department's own incident reporting form, if a fire incident was involved in the fatality; medical data such as death certificates or autopsy reports; special investigation reports from other agencies; police and motor vehicle accident reports, if applicable; photographs and diagrams; and additional newspaper accounts. Incidents to be included in the study are then coded into NFPA's Fire Incident Data Organization (FIDO) which includes both incident and casualty information. By mutual agreement of the USFA and NFPA project staff, the same inclusion criteria were used for the USFA study as are used in the NFPA study.

Work described to this point was done as part of NFPA's ongoing program of data collection and analysis in the area of fire fighter fatalities and was completed at no cost to FEMA.

E. Additional Data Collection Completed for the Contract

To meet FEMA's request for a list of the next-of-kin of the 1986 fatalities and the names and addresses of the fire chiefs, a follow-up mailing was sent to all departments asking them to verify the victims' names and dates of fatal injury, the names and addresses of the departments and chiefs, and the names and relationships of the next of kin. Telephone calls were made to non-responding fire departments to obtain the information.

II. 1986 FINDINGS

One hundred thirteen fire fighters died in the line of duty in 1986. As shown in Figure 1, although this was a 8.1 percent decrease from the year before, there has been little change over the past eight years, indicating that a plateau has been reached and a renewed effort must be made to target certain specific areas of the problem in order to achieve further reduction. A comparison of fire fighting to other occupations is shown in Figure 2. The data for other industries was obtained from a NIOSH report. The fire fighter fatality data was based on deaths of career fire fighters reported to NFPA from 1980 to 1984 and NFPA's estimate of the number of career fire fighters in the United States in 1983.[2] After mining, fire fighting is the most hazardous occupation in terms of fatalities. This study will report some of the most frequently occurring scenarios and will present some conclusions and recommendations to address the problem.

A. Type of Duty

The distribution of deaths by type of duty being performed is shown in Figure 3. The largest proportion of deaths occurred during fire ground operations. Of these 46 deaths, 26 were due to heart attacks, 13 to smoke inhalation, four to crushing injuries, two to internal trauma and one to stroke. Twenty-five of the victims were career fire fighters and 21 were volunteers.

The second largest category involved responding to and returning from alarms which accounted for one third of the deaths - the highest proportion over the past 10 years. Twenty-one of these 38 deaths were due to motor vehicle accidents, 13 to heart attacks, three to falls from apparatus, and one to a fall while responding on foot in a wooded area.

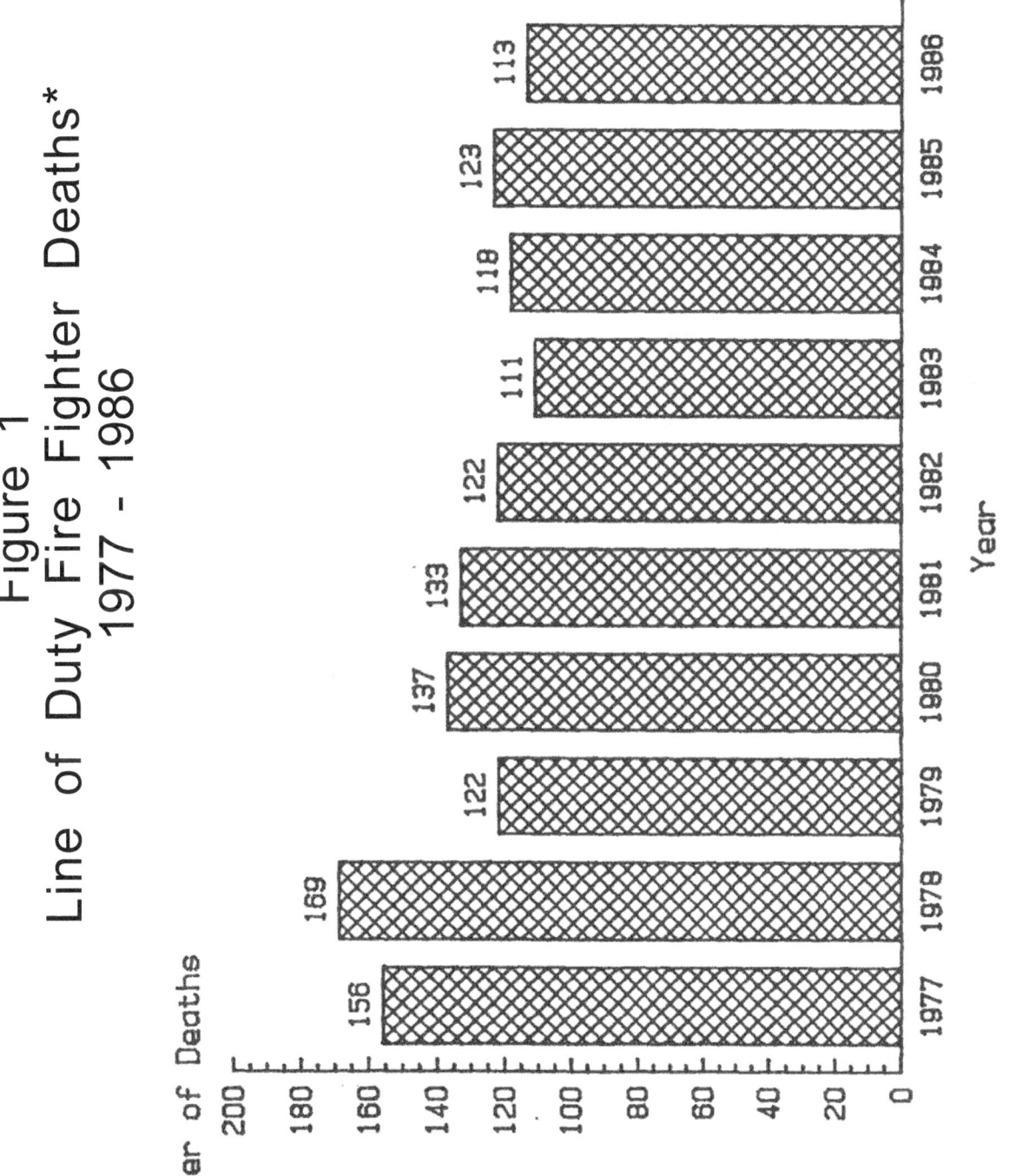

Figure 1
Line of Duty Fire Fighter Deaths*
1977 - 1986

* The number of fire fighter deaths may differ
from previously reported numbers because of
information received since publication.

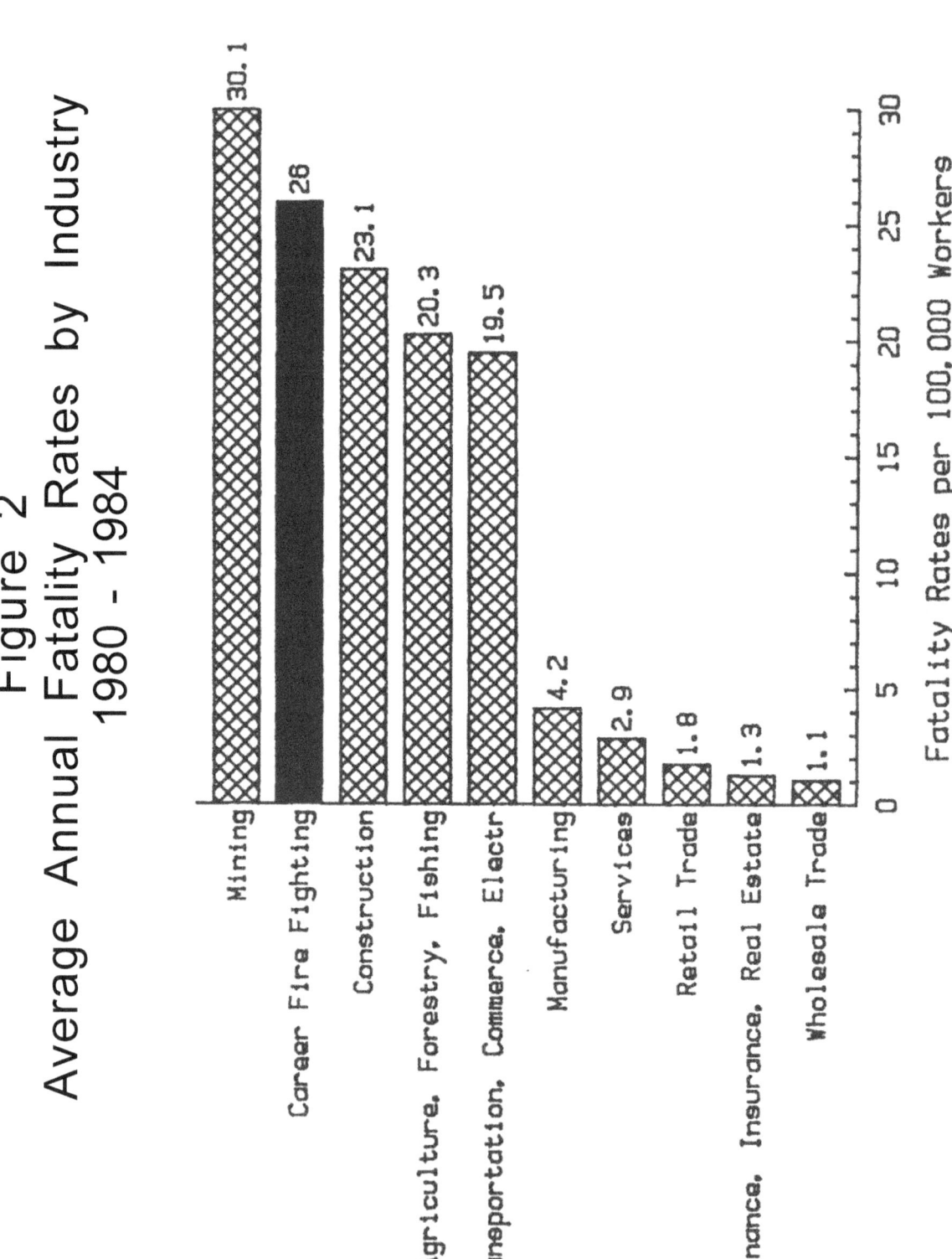

Figure 2
Average Annual Fatality Rates by Industry
1980 - 1984

Source: Fire fighter fatality data from
NFPA. All other data from NIOSH.

Figure 3
Fire Fighter Deaths 1986
by Type of Duty

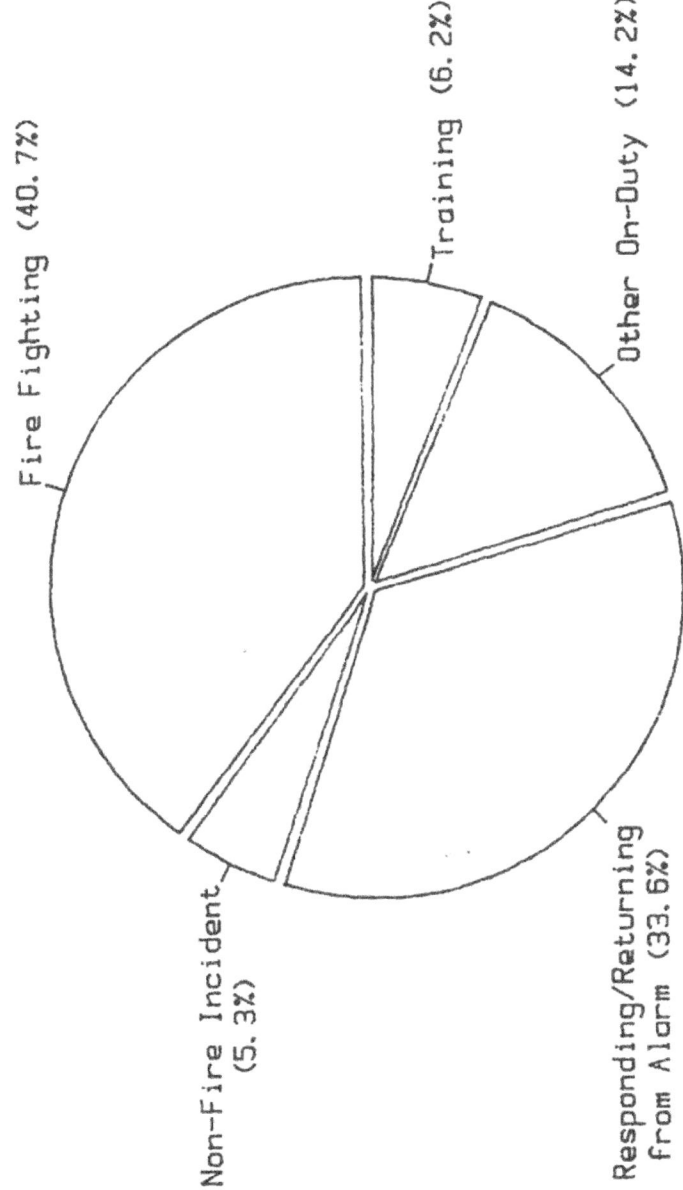

Fire Fighting (40.7%)

Training (6.2%)

Other On-Duty (14.2%)

Responding/Returning
from Alarm (33.6%)

Non-Fire Incident
(5.3%)

There were seven training-related fatalities - three heart attacks, two motor vehicle accidents, one stroke, and one fall from a ladder. There were six deaths while working at non-fire emergencies. These included two heart attacks during EMS calls and one during an ice rescue, one electrocution during an extrication of a person trapped in a motor vehicle accident, one victim struck by a train while evacuating a flooded area and one caught in an explosion while investigating a steam leak.

The other 16 deaths occurred while performing other duties - 10 deaths during normal station and administrative duties, two during fire inspections, one during maintenance, one supervising a fireworks display, one while on EMS standby and one while working as a fire policeman.

B. Cause and Nature of Fatal Injury or Illness

As used in this study, the term "cause" refers to the action, lack of action, or circumstances that directly resulted in the fatal injury while the term "nature" refers to the medical nature of the fatal injury or illness or what is often referred to as the cause of death. Often, the fatal injury is the result of a chain of events, the first of which is recorded as the cause. For example, if a fire fighter is struck by a collapsing wall, becomes trapped by the debris, runs out of air before being rescued and dies of asphyxiation, the cause of fatal injury recorded is "struck by collapsing wall" and the nature of fatal injury is "asphyxiation."

Figure 4 shows the distribution of deaths by cause of fatal injury or illness. Stress was reported as the cause in over half of the deaths. Nine of these 59 deaths were attributed to physical exertion on the fire ground or during training. The second major category was struck by or contact with objects. These 29 deaths included 24 motor vehicle accidents, three fire fighters struck by collapsing walls, one electrocution and one gunshot.

Figure 4
Fire Fighter Deaths 1986
by Cause of Injury

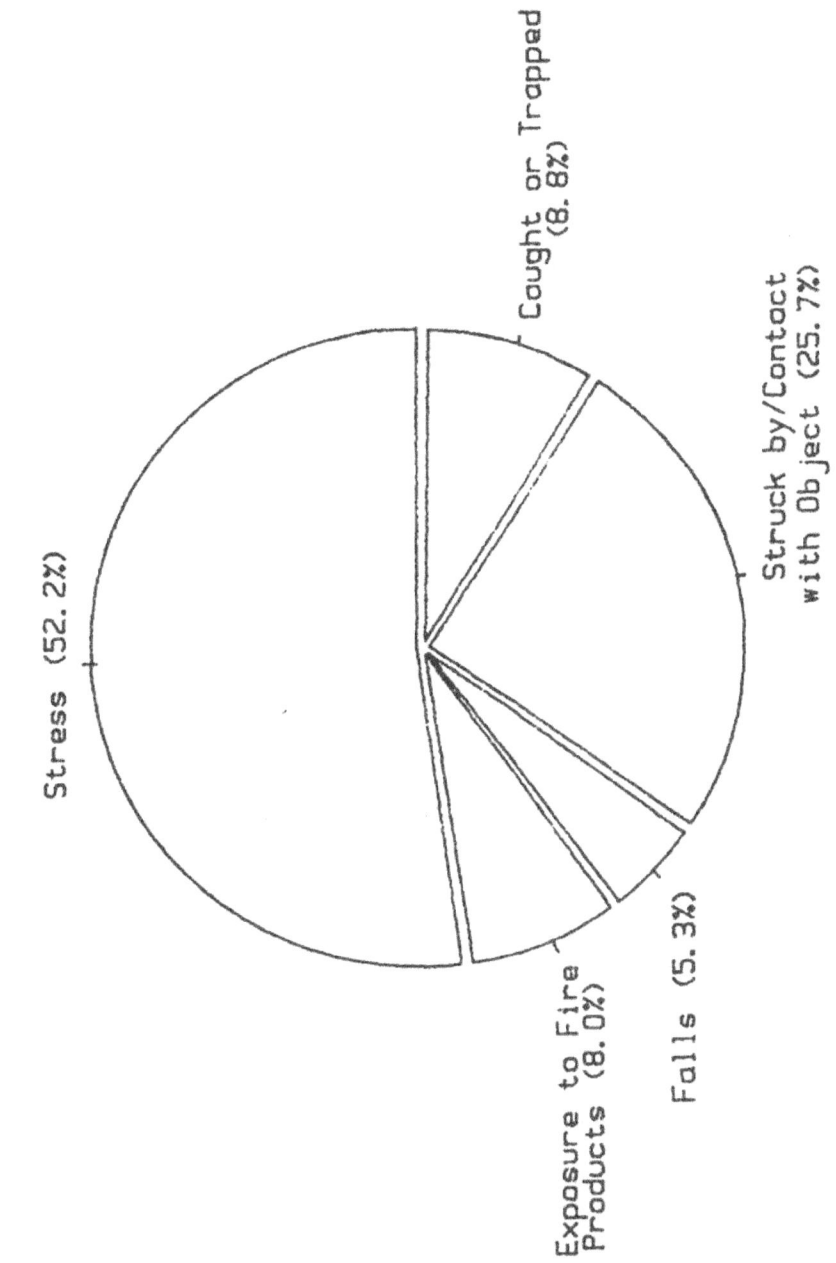

Stress (52.2%)

Caught or Trapped (8.8%)

Struck by/Contact with Object (25.7%)

Falls (5.3%)

Exposure to Fire Products (8.0%)

Ten fire fighters were caught or trapped - two each by interior collapse, rapid fire progress, explosions and falling objects, one by flashover and one was lost inside a building. Nine deaths were caused by exposure to fire products - eight to smoke and one to heat. Six fire fighters died as a result of falls - three from fire apparatus, one from a structure, one from a ground ladder and one tripped over a log while responding on foot in the woods. Fatal falls from apparatus over the past ten years are discussed in more detail later in this report.

Figure 5 shows the distribution of deaths by the medical nature of the fatal injury or illness. The largest proportion of deaths were due to heart attacks. Of these 58 deaths, medical documentation indicated that five of the victims had prior heart problems, either previous heart attacks or bypass surgery and 11 others had severe arteriosclerotic heart disease (defined for this study as arterial occlusion of at least 50 percent). Heart attack deaths are discussed in detail later in this report.

The other categories of nature of fatal injury were internal trauma (26 deaths), asphyxiation (14 deaths), crushing (9 deaths), stroke (3 deaths), and one each due to bleeding, electrocution and gunshot wound.

C. Ages of Fire Fighters

The ages of fire fighters who died in 1986 ranged from 18 to 75 years with a mean age of 44.2 years. The distribution of fire fighter deaths by age and cause of death is displayed in Figure 6. As might be expected, the proportion of heart attack deaths among older fire fighters is very high while traumatic injuries account for almost all of the deaths of fire fighters aged 40 and under. More than three-quarters of the fire fighters over 40 who died in 1986 suffered heart attacks. No fire fighter under the age of 36 suffered a fatal heart attack.

Figure 5
Fire Fighter Deaths 1986
by Nature of Injury

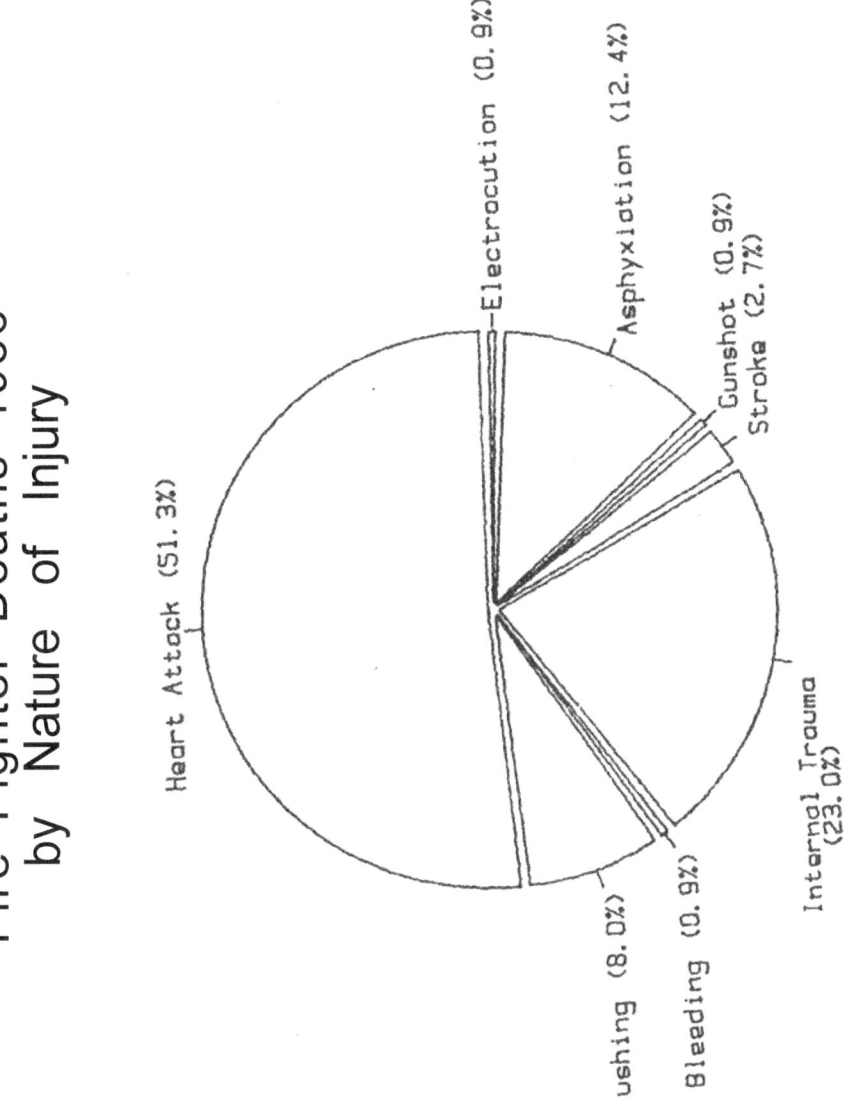

Electrocution (0.9%)

Asphyxiation (12.4%)

Gunshot (0.9%)
Stroke (2.7%)

Heart Attack (51.3%)

Internal Trauma
(23.0%)

Crushing (8.0%)

Bleeding (0.9%)

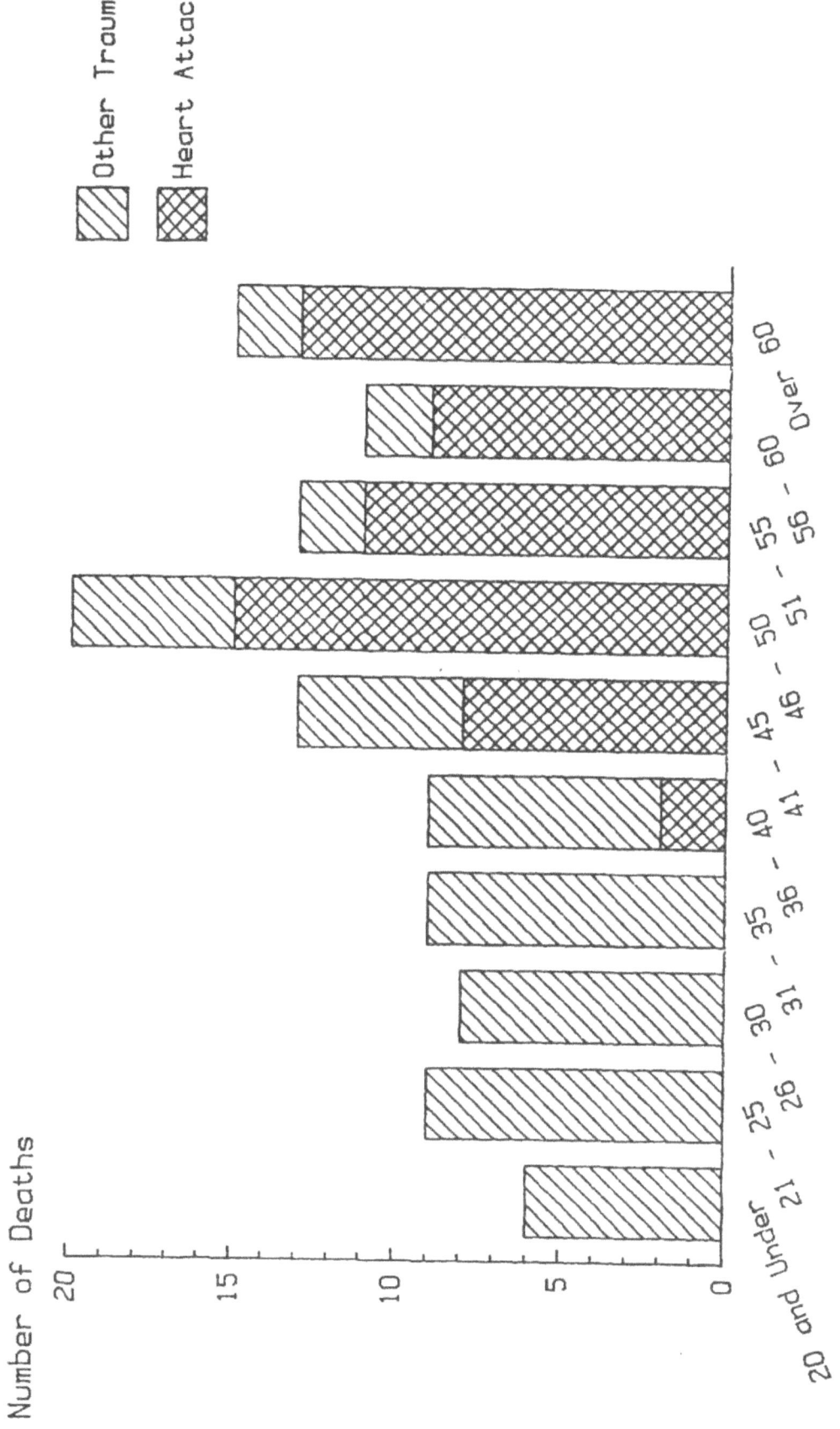

Figure 6
1986 Fire Fighter Deaths
by Age and Cause of Death

D. Fire Ground Deaths

The distribution of the 46 fire ground deaths by fixed property use is shown in Figure 7. The largest proportion of deaths occurred in residential properties. These 18 deaths included 11 in one- and two-family dwellings, six in apartment buildings and one in a residential hotel.

There were eight deaths in brush and forest fires. Four deaths each occurred in stores and public assembly properties, including two in a nightclub fire and one each in a restaurant and a church. Three deaths occurred in storage properties including one in a fire station. Two deaths were the result of fires in manufacturing properties. There was one death each at a nursing home, a dump, along a pier, and in a building under construction.

One fire fighter died of smoke inhalation in a fire of suspicious origin in a vacant building. One fire fighter suffered a heart attack at the scene of a truck fire on a freeway. Another fire fighter suffered a fatal heart attack while fighting a fire in a boat stored at a residence.

To put the hazards of structural fire fighting in perspective, the number of deaths per 100,000 structure fires was examined by fixed property use. The rates were calculated using the estimates of fire experience from NFPA's 1986 fire loss study.[3] There were 3.1 deaths per 100,000 residential structure fires in 1986 and 12.8 deaths per 100,000 nonresidential structure fires. Although almost three times as many fires occur in residential structures, the size, complexity and special hazards often associated with nonresidential structures result in a much greater risk at such fires.

Figure 7
Fire Ground Deaths in 1986
by Fixed Property Use*

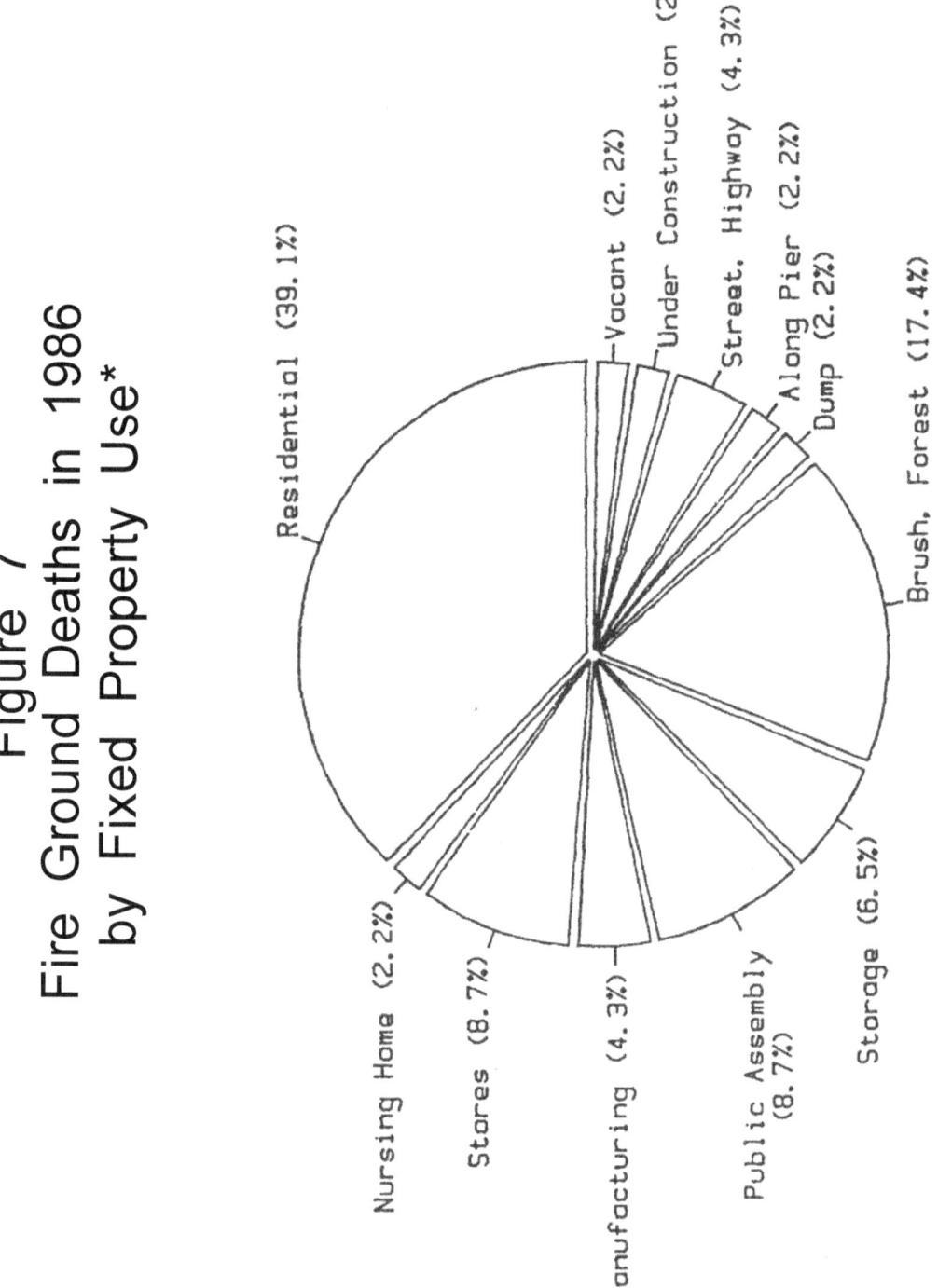

Residential (39.1%)

Vacant (2.2%)

Under Construction (2.2%)

Street, Highway (4.3%)

Along Pier (2.2%)

Dump (2.2%)

Brush, Forest (17.4%)

Storage (6.5%)

Public Assembly (8.7%)

Manufacturing (4.3%)

Stores (8.7%)

Nursing Home (2.2%)

* There were 46 fire ground deaths in 1986.

E. Time of Day

The distribution of 1986 fire ground deaths and total deaths by time of day is shown in Figure 8. This graph seems to indicate two peak times of day in both categories. For fire ground deaths, there is a first peak at mid- to late afternoon and then a higher peak between 1:00 and 3:00 am. The number of deaths overall peaked in the late afternoon to early evening, dropped off through the evening and peaked again between 1:00 and 3:00 am. In order to determine if these results for one year are significant, the distribution by time of day for a ten-year period was examined. These results are shown in Figure 9. The highest number of deaths both on the fire ground and overall occurred between 7:00 and 9:00 pm. The number of fire ground fatalities, however, is fairly constant between 1:00 and 9:00 pm. The lowest number of fire ground fatalities occurred between 5:00 and 11:00 am.

F. Month of the Year

Figure 10 shows the distribution of 1986 fire fighter deaths by month. The same information for 1977 through 1986 is shown in Figure 11. Nine of the 17 deaths in August 1986 were the result of three multiple-fatality incidents. There is no obvious pattern to the fire ground deaths in 1986. For the 10 year analysis, fire ground deaths are higher in the winter and in midsummer.

G. State and Region

The distribution of fire fighter deaths by state is shown in Table 1. Thirty-six states are represented on the list, led by New York with 10 deaths. The experience by region[4] is displayed in Table 2. The South lost the highest number of fire fighters (39), followed by the Northeast (35), the Northcentral region (22) and the West (17). When looking at fire ground deaths we see that although there were more fire ground deaths in the South; there were so many more fires in the South than in the Northeast that the death rate per 100,000 fires was highest in the Northeast.

-17-

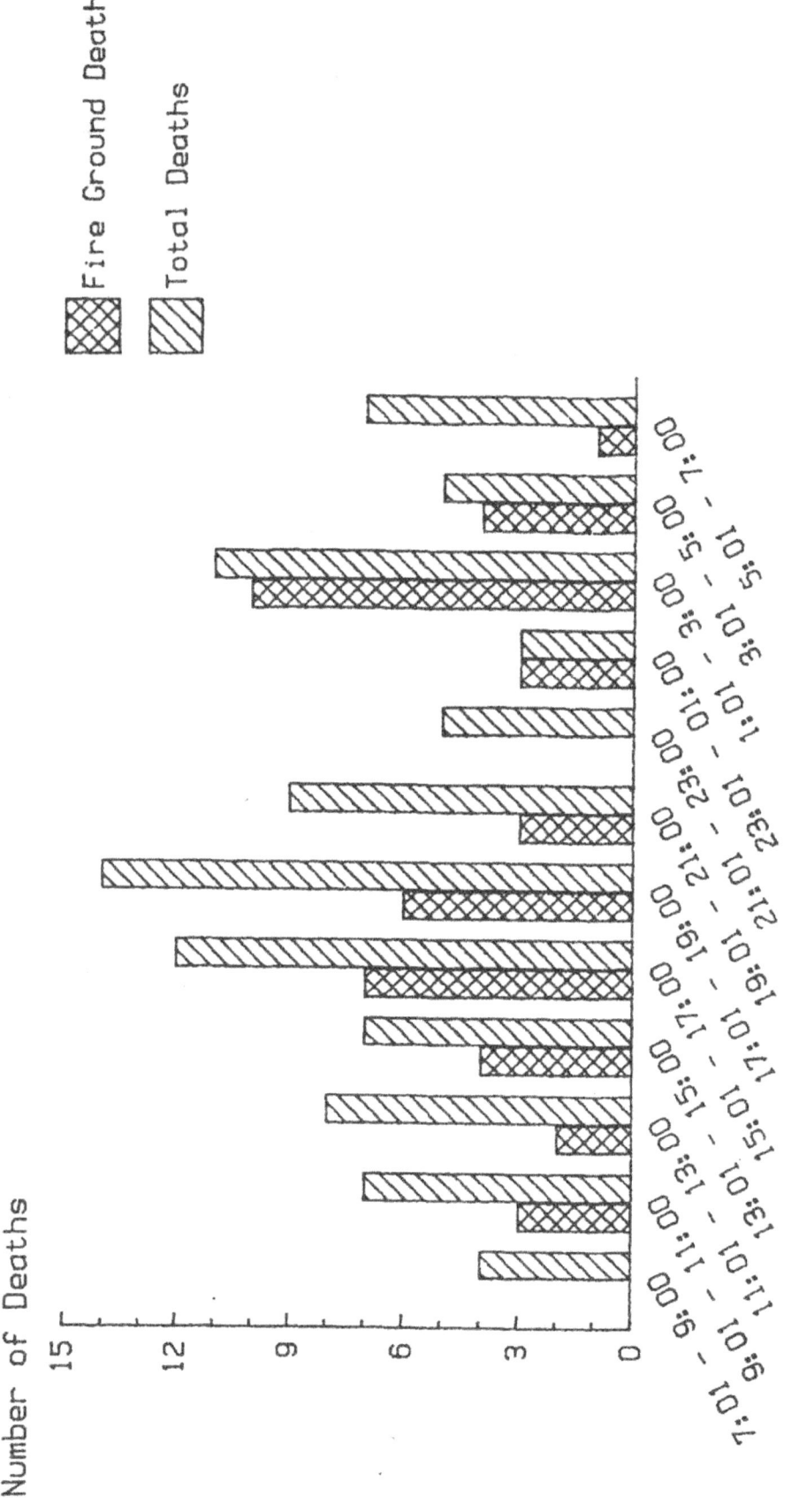

Figure 8
Fire Fighter Fatalities by Time of Day
1986

Fire Ground Deaths

Total Deaths

Number of Deaths

Time of Day

Based on 43 fire ground fatalities and
92 total fatalities for which time
was known.

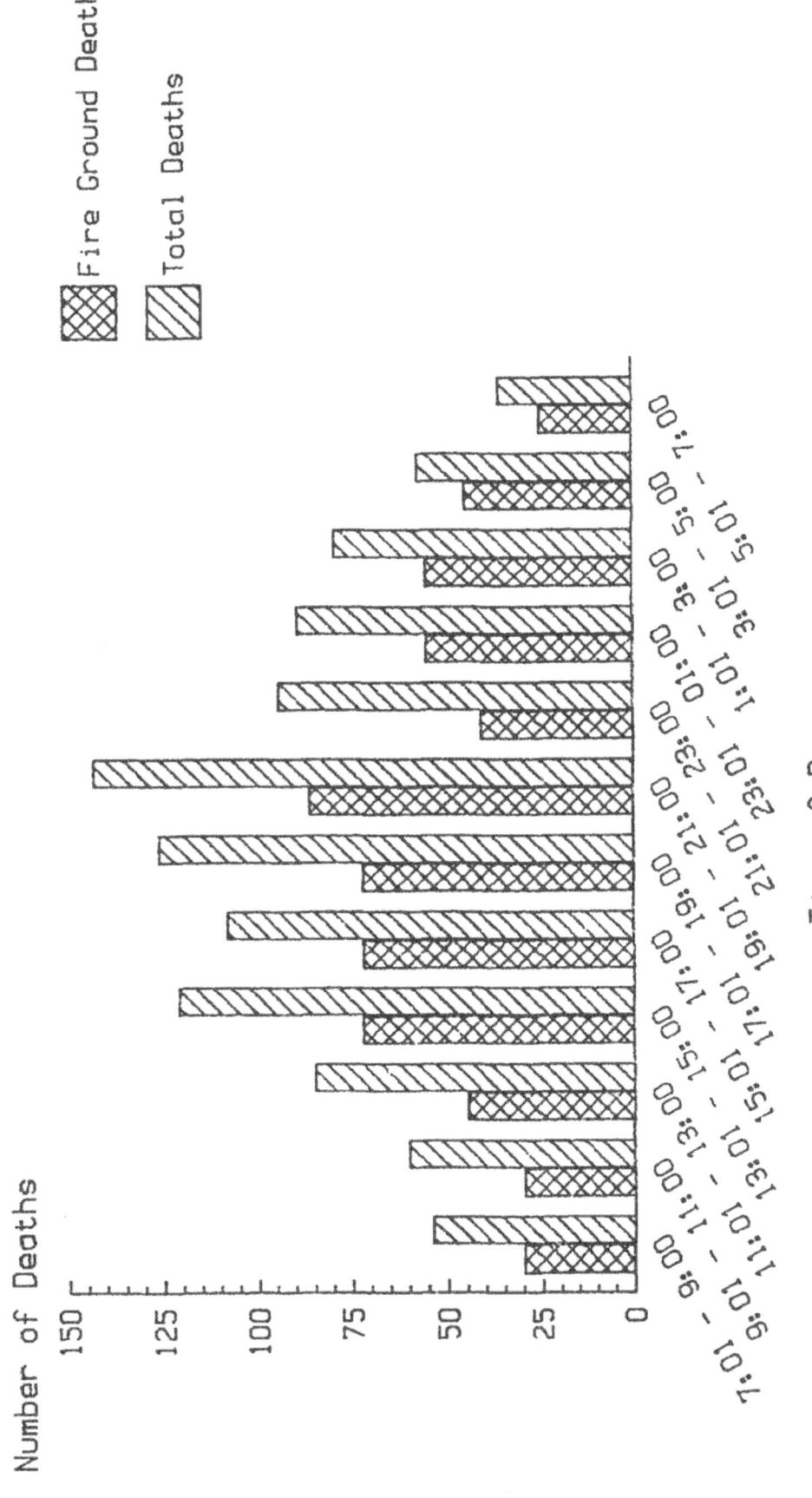

Figure 9
Fire Fighter Fatalities by Time of Day
1977 - 1986

Based on 628 fire ground fatalities and
1052 total fatalities for which time
was known.

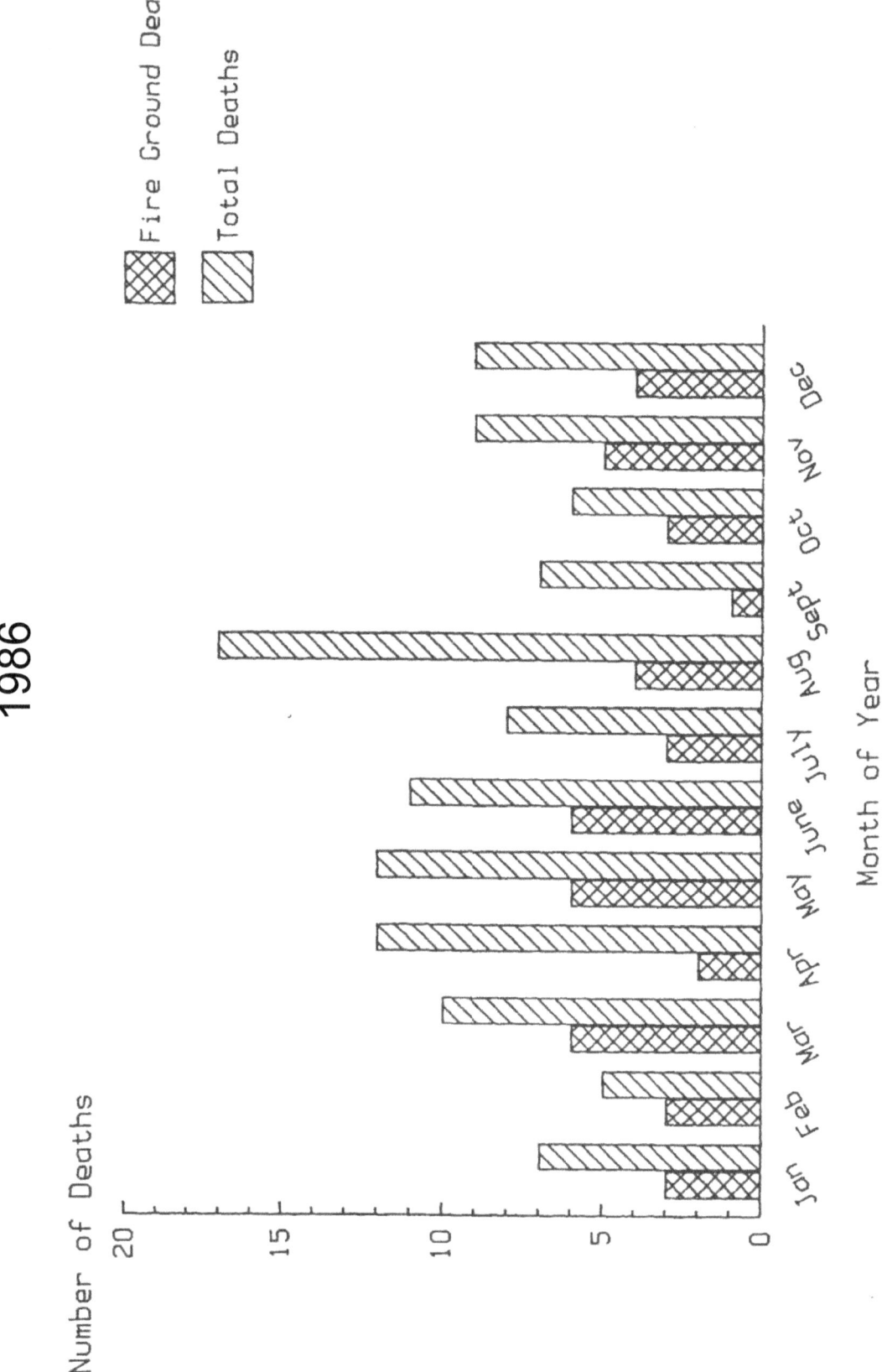

Figure 10
Fire Fighter Fatalities by Month of Year
1986

Fire Ground Deaths

Total Deaths

Number of Deaths

Month of Year

Based on 46 fire ground fatalities and
113 total fatalities.

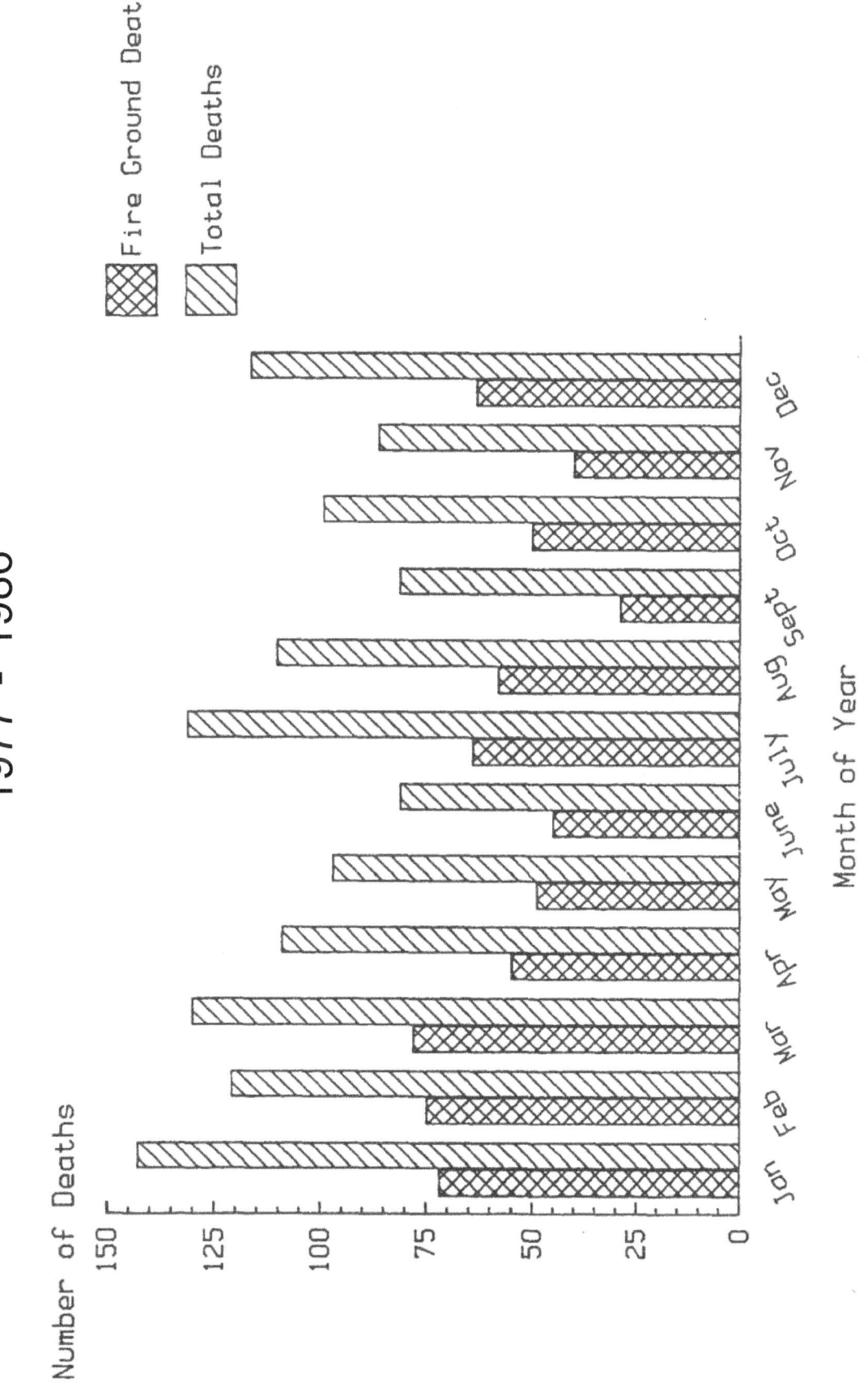

Figure 11
Fire Fighter Fatalities by Month of Year
1977 - 1986

Based on 678 fire ground fatalities and
1304 total fatalities.

Table 1
1986 Line of Duty
Fire Fighter Fatalities

State	Number of Deaths	State	Number of Deaths
Alabama	1	Mississippi	2
Arizona	1	Missouri	2
Arkansas	2	Nevada	1
California	2	New Jersey	7
Colorado	4	New York	10
Connecticut	4	North Carolina	4
Florida	4	Ohio	7
Georgia	3	Oklahoma	1
Idaho	6	Oregon	1
Illinois	4	Pennsylvania	8
Indiana	3	Rhode Island	1
Kansas	2	South Carolina	4
Kentucky	3	Tennessee	2
Louisiana	2	Texas	5
Maryland	2	Virginia	2
Massachusetts	5	Washington	2
Michigan	1	West Virginia	2
Minnesota	1	Wisconsin	2

TOTAL: 113

Table 2
Fire Fighter Death Rate by Region
1986

Region	Number of Fatalities	Number of Fire Ground Deaths	Fire Ground Death Rate per 100,000 Fires
Northeast	35	15	2.82
Northcentral	22	9	1.84
South	39	19	2.15
West	17	3	0.82
Nationwide	113	46	2.03

H. Analysis of Urban/Rural/Suburban Patterns in Fire Fighter Fatalities

The U.S. Bureau of the Census defines "urban" as a place having at least 2,500 population or lying within a designated urbanized area. "Rural" is defined as any community that is not urban. "Suburban" is not a Census term but may be taken to refer to any place, urban or rural, that lies within a metropolitan area defined by the Census but is not one of the designated central cities of that metropolitan area.

Fire department coverage areas do not always conform to the boundaries of Census places. For example, fire departments defined by counties or special fire protection districts may have both urban and rural sections, and there are Federal, state, and private fire fighters. In such cases, it may not be possible to characterize the entire coverage area of a fire department as rural or urban, and one must assign a fire fighter death as urban or rural based on the particular community in which he was operating when fatally injured.

Based on these rules, the following patterns were found and are shown with available patterns for the general population and for the population of fire fighters specifically in local fire departments:

	Urban	Rural	Total
Total 1986 fire fighter fatalities	64 (57%)	49 (43%)	113 (100%)
Suburban location	26	9	35
Local fire department only*	62 (62%)	38 (38%)	100 (100%)
U.S. population (1980)	74%	26%	100%
U.S. fire fighters (1985), total**	59%	41%	100%
U.S. fire fighters (1985), career**	98%	2%	100%
U.S. fire fighters (1985), volun.**	47%	53%	100%

---

* Excludes one military service fire fighter and one private company fire fighter killed in urban locations and 11 Federal and state fire fighters killed In rural locations.

** "Estimated Fire Fighters in the U.S., 1985," Quincy, Massachusetts: National Fire Protection Association, Fire Analysis Division, February 1987. All percentages are for fire fighters in local fire departments.

The distribution of fire fighter fatalities from local fire departments (62% urban, 38% rural) is closer to the distribution of fire fighters from local fire departments (59% urban, 41% rural) than to the distribution of the whole U.S. population (74% urban, 26% rural).* This suggests that the risk of dying, as measured by local fire fighter deaths per 100,000 local fire fighters,is nearly the same in urban and rural areas.Urban areas have a considerably lower rate of local fire fighter deaths per million persons protected than do rural areas, but that appears to be because urban areas have fewer fire fighters per million persons protected.

This finding of roughly equal risk of death per 100,000 fire fighters for urban and rural areas is surprising, however, because rural areas are served (as noted> almost entirely by volunteer fire fighters, who average fewer work hours per year than career fire fighters and who would therefore be expected to have a lower risk of death on a per 100,000 fire fighters basis. In fact, volunteers had one-third of the fire fighter death rate of career fire fighters in 1986. (Volunteers accounted for 53 of the 100 local fire fighters, or 53%. versus 78% of the local fire fighters, according to 1985 figures.Therefore, the risk index for volunteers would be .53/.78=.68 versus .47/.22=2.14 for career fire fighters,the latter being three times the former.>

Therefore the urban/rural comparison gives results quite different from the career/volunteer comparison. Part of this difference is due to the fact that urban and rural are defined somewhat differently for fire fighters then for fire fighter deaths (as noted in the footnote). The rest of the

---

* Note that the classification of fire fighters into urban and rural is based strictly on the population protected by the fire department and not on metropolitan area considerations. However, if fire fighter fatalities were similarly classified, the distribution would shift by at most two percentage points,so the points here are not affected.

difference apparently reflects a much lower death rate for urban volunteers then for either urban career fire fighters or for rural volunteers. (There are no rural career fire fighters to speak of.) This point would be worth rechecking for the future using multi-year data.

I. Additional Analysis

In an effort to provide a means of alerting fire fighters to incidents with particularly dangerous potential, various theories were explored, including some that might be regarded very skeptically. One such analysis looked at the use of biorhythms to predict times requiring extra caution. The theory and analysis are discussed below, but the results indicated that biorhythms do not influence fire fighter fatalities.

The theory of biorhythms postulates, among other things, a relationship between frequency of accidents and three cycles affecting lives.* Each cycle is represented by a sinusoidal wave of fixed periodicity. The three waves are the physical cycle (23-day period), the emotional cycle (28-day period), and the intellectual cycle (33-day period). All three cycles are postulated to begin at their baseline values on the day of birth and to rise initially. With this information, it is possible to calculate biorhythm values. for any person for any day, given only that person's date of birth.

The theory identifies any day in which one or more cycles crosses the baseline as a critical day. This transitional phase between peak and valley is believed to be associated with a higher probability of accidental injury or death. Each cycle crosses the baseline twice during its period. However, two of the three cycles have periods measured by an odd number of days. For example, in plotting the physical cycle, the first critical day after birth

---

*   See, for example, Bernard Gittelson, Bio-Rhythm -- A Personal Science, New York: Warner Books, Inc., 1977.

should occur 11 1/2 days after birth. The interpretation or this main day within the theory is not completely clear. It may mean that there are two "half critical" days or two fully critical days where one would expect only one.The analyses here have allowed for both possibilities.

The probability that any cycle is on a critical day is equal to one minus the probablity that no cycle is on a critical day. Because the three cycles have period lengths with no common factors (23 is prime, 28= 2x2x7= 4x7, 33= 3x11), the probability that any day is critical on one cycle is independent of whether that day is critical on any other cycle. Accordingly, the probability that any cycle is on a critical day is given by the following:

1. If half-critical days are counted as half-critical, then each cycle has two critical days per period, and the probability formula is:

$$1 - (21/23) \ (26/28) \ (31/33) = 0.204$$

2. If half-critical days are counted as fully critical, then the physical and intellectual cycles have three critical days per period, and the probability formula is:

$$1 - (20/23) \ (26/28) \ (30/33) = 0.266$$

Of the 113 fire fighter fatalities in 1986, 67 had known dates of birth. The 67 deaths included 7 occurring on fully critical days and 8 occurring on half-critical days. If half-critical days are counted as half-critical, then the 67 deaths included 7 + (l/2)(8)=11 critical days, compared to (0.204)(67)=13.6 predicted critical days. If half-critical days are counted as fully critical, then the 67 deaths included 7+8=15 critical days, compared to (0.266)(67)=17.8 predicted critical days.

In both analyses, the actual number of critical days was slightly less then predicted, indicating a lower risk of fatal injury on critical days, but not lower by a statistically significant margin. This result indicates the theory of biorhythms is not confirmed as influencing fire fighter fatalities.

III. FATAL HEART ATTACKS 1977-1986

Over the past 10 years, heart attack has been, the leading cause of fire fighter fatalities, accounting for almost half of the total deaths. This problem was first examined in a special study published by NFPA[5]. The following analysis is based on the updated file of fire fighter deaths.

For this analysis, the term "heart attack" was expanded to include cardiac arrest, stroke (CVA) and aneurysm, because all involve the cardiovascular system. Of the 1304 fire fighter fatalities over the 10-year period, 614 (47.1 percent) were due to heart attacks -588 cardiac arrests, 21 strokes and 5 aneurysms. This study focuses on those 614 deaths. The aspects of the heart attack problem examined in this study include the physical condition of the victims prior to their deaths,the types of duty they were engaged in and their ranks and ages.

A. Physical Condition

For 259 of the 614 heart attack victims, information on their physical conditions prior to their fatal injury was available. This information was obtained from medical documentation that accompanied fatality reports.* As shown in Figure 12, of these 259 fire fighters, 105 (40.5 percent) had had some prior heart-related conditions, such as previous heart attacks or

* Medical documentation used to evaluate the physical conditions of heart attack victims was in the form of death certificates and autopsy reports made available by PSOB and fire departments. Fire fighters who experienced prior heart problems may or may not have been aware of their physical condition prior to the fatal heart attack.

Figure 12
Fire Fighter Heart Attack Deaths
by Physical Condition 1977 - 1986

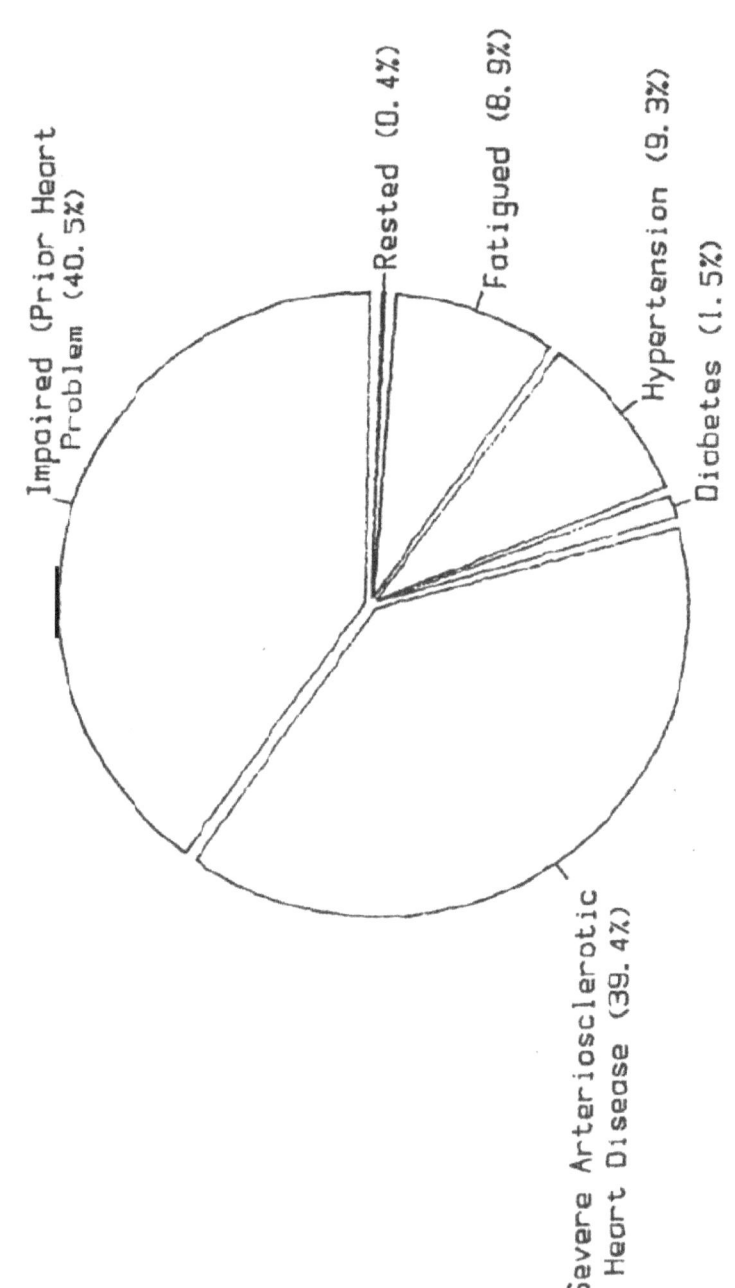

Impaired (Prior Heart Problem (40.5%)

Rested (0.4%)

Fatigued (8.9%)

Hypertension (9.3%)

Diabetes (1.5%)

Severe Arteriosclerotic Heart Disease (39.4%)

coronary bypass surgery. Another 102 or the victims (39..4 percent, now severe arteriosclerotic heart disease (defined for this study as medically documented arterial occlusion of at least 50 percent). Taken together, then, 80 percent of the heart attack victims for whom medical documents were available had known or detectable heart problems and were still active in the fire service.

Hypertension - another easily diagnosed, serious condition - was indicated in 24 (9.3 percent) of the victims and four others (1.6 percent) had diabetes. Only 24 of the 259 victims had no previous health problems - 23 of them were reported to have been fatigued.

Available reports on the other 355 heart attack victims did not contain enough medical documentation to determine whether previous, related problems existed. Therefore, it is possible that the number of victims with previous heart conditions may be considerably higher than the 207 cited above.

B. Type of Duty

Figure 13 shows the distribution of types of duty the 614 fire fighters were engaged in when they suffered their fatal heart attacks. The largest proportion, 306 deaths (49.8 percent), occurred during fire ground activities. The next largest proportion involved fire fighters responding to or returning from alarms (155 deaths or 25.2 percent). The other major categories included 70 deaths during administrative and normal station duties (11.4 percent), 35 deaths during training (5.7 percent) and 34 deaths while working at non-fire incidents such as emergency medical and rescue calls (5.5 percent). The other 14 deaths (2.3 percent) occurred during other on-duty activities such as fire prevention, inspection, maintenance, etc.

Figure 13
Fire Fighter Heart Attack Oeaths
by Type of Duty 1977 - 1986

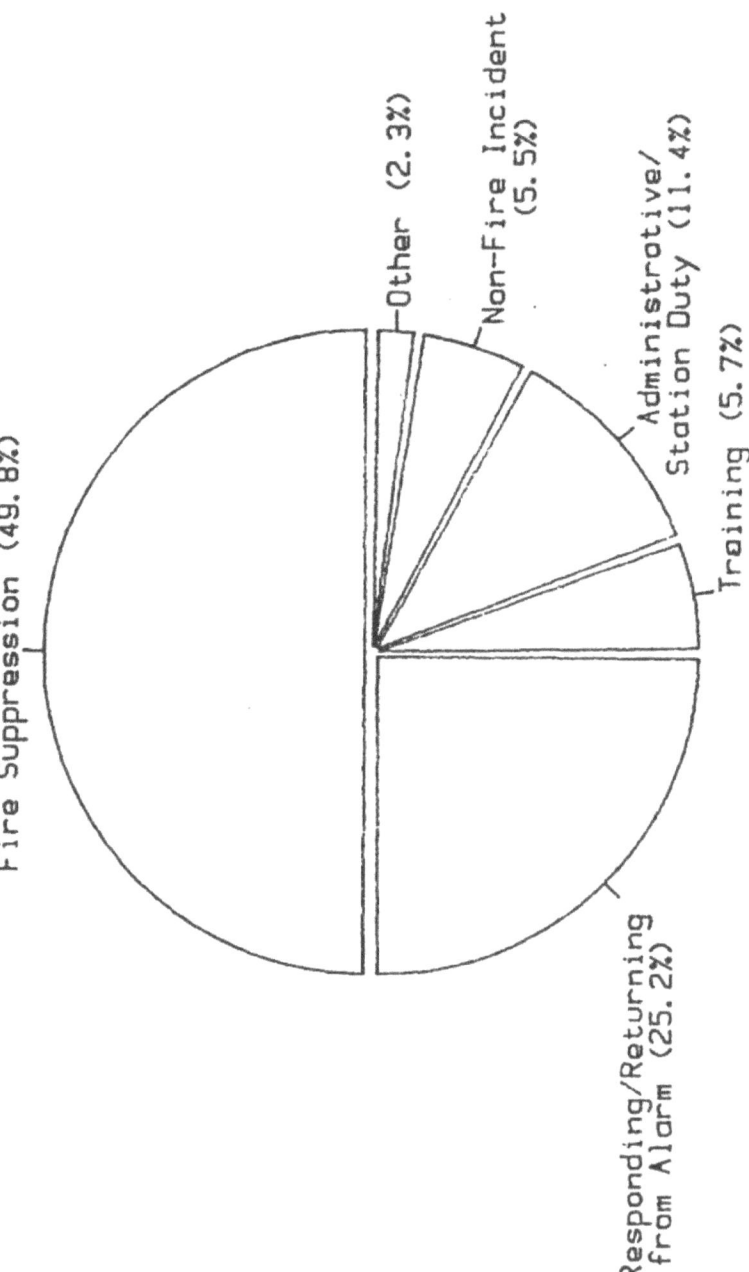

Other (2. 3%)

Non-Fire Incident
(5. 5%)

Administrative/
Station Duty (11. 4%)

Training (5. 7%)

Fire Suppression (49. 8%)

Responding/Returning
from Alarm (25. 2%)

C. Ages of Fire Fighters

The distribution of ages of the heart attack victims is displayed in Figure 14. The ages range from 20 to 87 years, with a mean age of 51.5 years. As can be seen, the breakdown by career and volunteer is fairly equal (334 career fire fighters vs. 280 volunteer fire fighters) although it is estimated that only 22 percent of the nation's fire fighters are career. Therefore, career fire fighters suffer proportionally more heart attack deaths than volunteer fire fighters. (Career fire fighters in fact suffer proportionally more deaths of all types than volunteers, probably because of much high number of hours and calls per year per person for career fire fighters.) Figure 14 also shows that the victims over 60 were mostly volunteers, possible due to the tendency of volunteer fire fighters to remain active well beyond the retirement ages that govern career fire fighters.

D. Ranks of Fire Fighters

Figure 15 shows the ranks held by the 609 fire service personnel who suffered fatal heart attacks and whose ranks were known. The largest proportion (60.6 percent) involved fire fighters. Company officers made up 23.0 percent of the total and the remaining 16.4 percent were chief officers. Based on these 609 fatalities, statistics indicate that, for every 2.6 fire fighters stricken with a fatal heart attack, one company officer is stricken. Based on typical complements of three or four personnel to each fire company (one officer and two or three fire fighters) this suggests a roughly equal risk of heart attack for company officers and fire fighters. However, for every 3.7 fire fighters stricken, one chief officer died, showing a much higher risk for chiefs (although the ratio of chief officers to fire fighters is not known). However, because fire service personnel at all levels face increased risk of heart attack, prevention programs should not focus on any one rank.

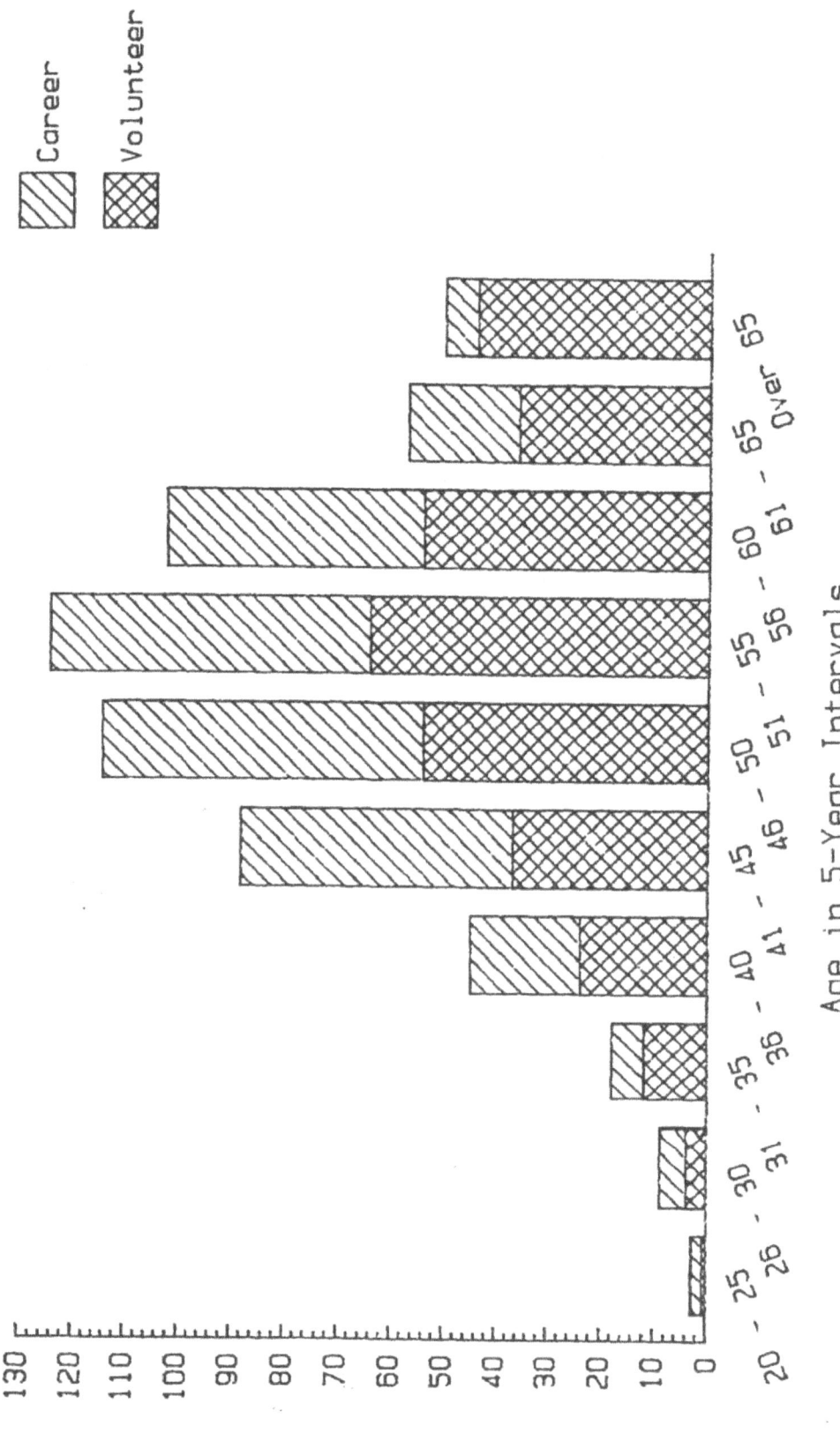

Figure 14
Fire Fighter Heart Attack Deaths
by Age 1977 - 1986

Figure 15
Fire Fighter Heart Attack Deaths
by Rank 1977 - 1986

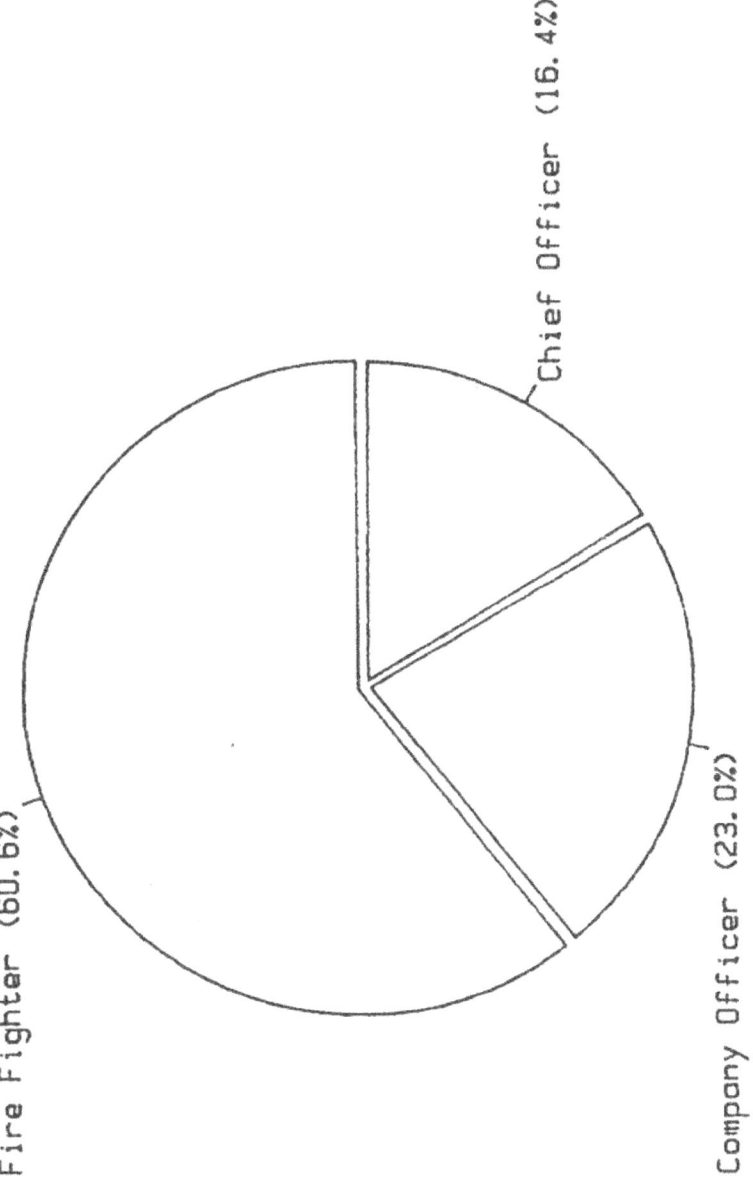

Fire Fighter (60.6%)

Chief Officer (16.4%)

Company Officer (23.0%)

E. Conclusion

From 1977 through 1986, 614 fire fighters, or almost half of the total number of fire fighters who died in the line of duty, have died as the result of heart attacks, and 80 percent of those with medically documented prior conditions had known or detectable heart problemsHeart attacks among fire service personnel are serious, all-too-common occurrences.Steps to reduce the risk of heart attacks among fire fighters, along with more detailed medical evaluations, should be taken. Further study into the areas of physical fitness and dietary requirements for fire fighters is recommended. Above all, attention must be focused on the significant problem of fire service personnel who have heart problems, yet are allowed to remain active in fire fighting.

IV. FATAL FALLS FROM APPARAIUS 1977 - 1986

In the 10-year period from 1977 through 1986, 50 fire fighters died as a result of falls from fire apparatus. This number does not include incidents where fire fighters fell or were thrown after impact in a motor vehicle accident and in no case was excessive speed a factor. Twenty-six of the victims were career fire fighters and 24 were volunteers.

Engines or pumpers were the vehicles most frequently involved in these incidents. In 19 of the 43 incidents involving engines, the victim fell from the backstep. In 17 incidents the victim fell from the jumpseat area, and in the other 7 cases the position of the victim was unknown. Tankers were involved in three incidents, ladders were involved in two and some other type of apparatus was involved in the other two. As would be expected, most of the injuries were internal trauma and crushing.

Figure 16 shows the major scenarios in falls from apparatus. In 18 of the incidents fire fighters lost their balance while riding apparatus, but not while the vehicle was making a turn. In these cases, they were often putting on their equipment or were thrown off when the vehicle suddenly stopped or accelerated or hit a bump in the road. In 17 incidents, fire fighters fell on turns. In these cases, they were most frequently donning protective coats or SCBA, reaching for equipment or signalling to other fire fighters. In seven incidents, fire fighters were fatally injured when they were struck after stepping or falling off vehicles still being positioned on the fire ground. In the other incidents, one fire fighter was killed when he hit or was hit by a malfunctioning station door; one fell from the cab of the vehicle he was driving when the door suddenly opened and he reached out to close it; and the third slipped or fell while getting out of the cab, was run over by the rear wheels and died later as a result of complications from his injuries.

Figure 16
Major Scenarios
in Fatal Falls from Apparatus

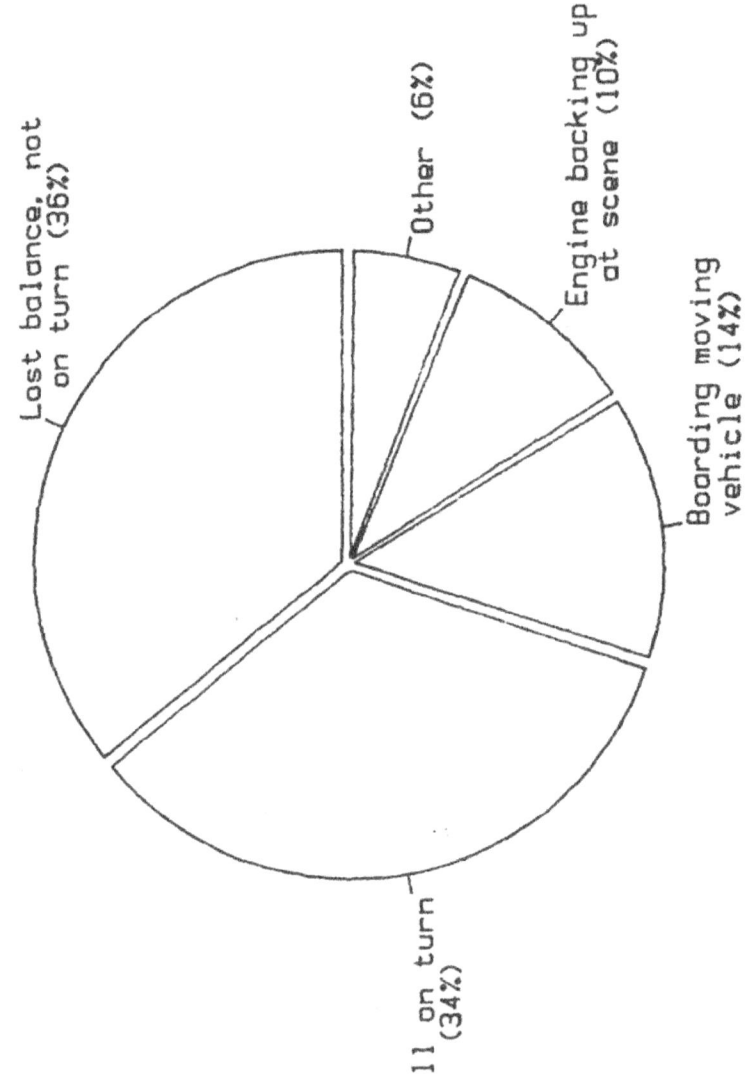

Lost balance, not on turn (36%)

Other (6%)

Engine backing up at scene (10%)

Boarding moving vehicle (14%)

Fall on turn (34%)

In order to reduce the frequency of such preventable fatal injuries, it is important to adopt the following safety precautions:

* Fire fighters must don protective clothing or equipment before the apparatus is in motion.

* Fire fighters must be seated, with seatbelts fastened, before the apparatus moves.

* Fire fighters must be prohibited from riding the sides and backsteps of apparatus.

* Seatbelts must be provided and used for all positions.

* Backup alarms should be part of apparatus safety equipment, and a backup person must be used whenever apparatus is backing.

* Fire fighters must be prohibited from boarding moving apparatus.

* Fire fighters must not dismount apparatus until it is stopped and the order is given.

* Apparatus operators must consider the safety of all persons in the immediate vicinity prior to moving apparatus.

* Fire officers must assume responsibility for the readiness of fire fighters prior to apparatus being moved to or from any location.

V. FATAL TANKER ACCIDENTS 1977-1986

From 1977 through 1986, 200 of the 1304 fire fighters fatalities (15.3 percent) were the result of motor vehicle accidents and almost one-fifth of those deaths involved tankers (see Figure 17). These 38 deaths were the result of 33 accidents. Thirty-three of the victims were killed while responding to alarms. Thirty-three were members of volunteer fire departments, a reflection of the greater likelihood of such departments using that type of fire apparatus.

Twenty-six of the accidents were attributed to failure to negotiate curves and losing control of the vehicle. Four of the deaths were due to heart attacks, one to burns and the remainder to traumatic injuries such as internal trauma and crushing.

Of the 33 tankers involved in these incidents, 11 were known to be converted former military or other vehicles, four were built for fire department use as tankers and no information was available on the other 18 vehicles. One of the converted vehicles had originally been used as an oil truck. Details on baffling were not generally available.

A review of road factors showed that 25 of the accidents occurred on paved roads and 21 occurred on curves or turns. Excessive speed was cited as a factor in 12 of the accidents.

The use of tankers is essential in rural fire fighting but extra care must be taken to consider their stability. Military vehicles have a high center of gravity as originally designed. Adding a tank above increases the vehicle's instability at higher speeds and on turns, and the weight of the water can seriously tax the truck's suspension, brakes and tires.

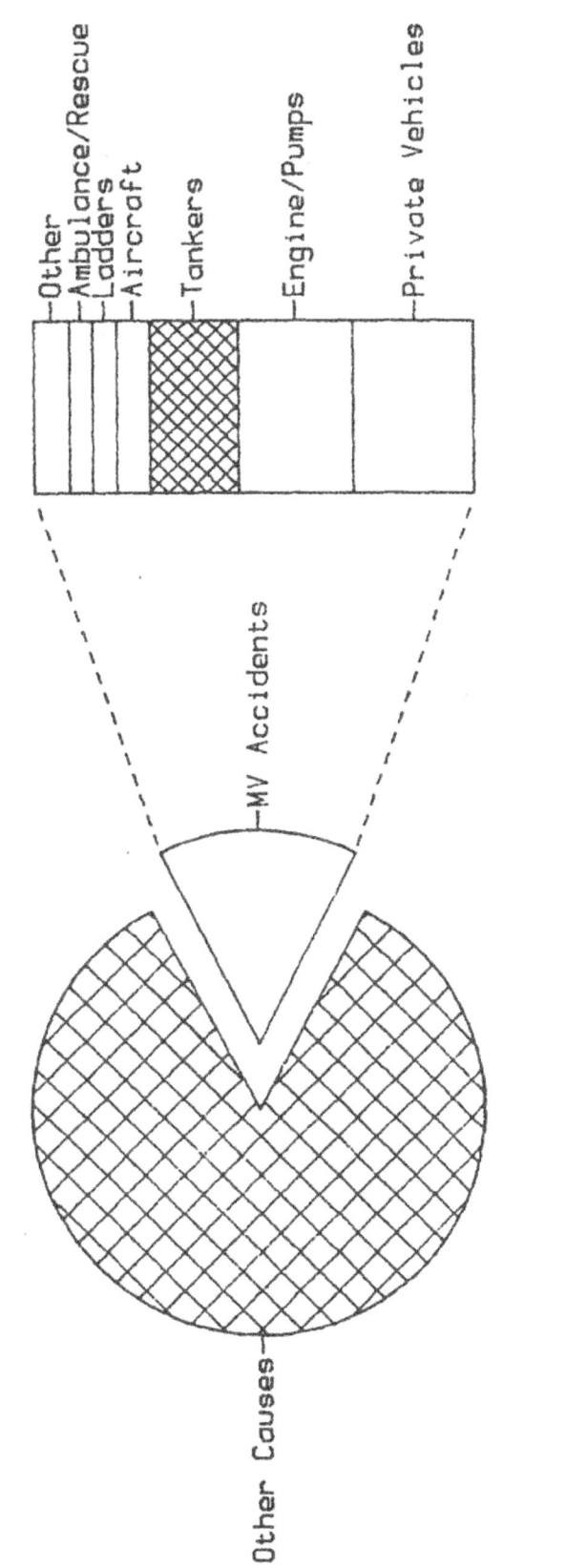

Figure 17
Motor Vehicle Accidents Resulting in
Fire Fighter Fatalities 1977 - 1986

Other
Ambulance/Rescue
Ladders
Aircraft
Tankers
Engine/Pumps
Private Vehicles

Types of Vehicles

MV Accidents

Other Causes

Cause of Deaths

The inexperience of drivers also played a role in the incidence of fatal tanker accidents. Inexperienced drivers include not only young people but also older fire fighters who are infrequently called upon to drive a vehicle that requires handling quite different from a passenger car or truck. Other factors included driver fatigue, driving while intoxicated, lack of safety equipment, and failure to observe traffic laws.

Increased training and an awareness of the potential problems with tankers are essential to reducing the frequency of deaths in tanker accidents. The major problems identified in accidents over the past 10 years included:

* inexperienced drivers

* speed of vehicles

    - excessive for vehicle type

    - excessive for road conditions

    - excessive for weather conditions

* excessive gross vehicle weight in converted vehicles

* high center of gravity of the vehicle/water tanks

* lack of use of safety equipment

* physical condition of driver

    - heart attacks

    - fatigued

    - driving while intoxicated

* failure to observe traffic laws.

VI CONCLUSIONS AND RECOMMENDATIONS

Although there has been some reduction in the number of fire fighter fatalities since the late 1970's. a plateau seems to have been reached and more aggressive efforts must be made to reduce fatalities further. The purpose of this report is to document some of the findings from the 1986 experience and the last decade's experience in order to use the lessons learned to improve fire fighter safety.

This report analyzed three areas of particular concern: heart attacks (the number one cause of fire fighter fatalities), and falls from apparatus and tanker accidents (two of the more preventable causes of death). The heart attack study showed the extremely high incidence of serious known or detectable heart problems among victims. Steps must be taken to reduce the risk of heart attacks among fire fighters including further study of the physical fitness and dietary requirements of fire fighters. A closer medical evaluation of fire fighters, particularly those who have experienced prior heart problems, is needed to exclude those personnel from active fire fighting roles.

Since 1977, slightly more than one quarter of the 1304 fire fighter deaths occurred while responding to or returning from alarms. In some years the proportion has been as high as one third. Most of these deaths occurred while responding. Fire fighting is a hazardous enough profession without such a high loss of life while trying to reach the emergency scene. Two of the frequent causes of fatalities - falls from apparatus and tanker accidents - were highlighted in this study, and steps that would lessen their occurrence were discussed earlier.

Areas that merit additional study include investigation into the medical histories of the 355 heart attack victims for whom documentation of prior heart problems was not available; an examination of the cancer deaths reported to PSOB, whether they qualified for benefits or not; and a comprehensive analysis of factors in motor vehicle accidents while responding to or returning from alarms. Two relevant topics in the latter category are fatal accidents involving volunteer fire fighters responding in their own vehicles and fatal and non-fatal accidents involving emergency vehicles.

To an individual fire department, the death of a fire fighter can appear to be a random and extremely rare event. However, a look at the national experience can provide valuable lessons to all departments. Changes in operating procedures and attitudes must be made to improve fire fighter safety.

REFERENCES

1. "National Traumatic Occupational Fatalities, 1980-1984," National Institute for Occupational Safety and Health, Division of Safety Research, Morgantown, WV, June 11, 1987.

2. Michael J. Karter, Jr., "Taking the Measure of the Fire Service," Fire Command, Vol. 52, No. 7, (July 1985).

3. Michael J. Karter, Jr., "A Look at Fire Loss in the United States During 1986," Fire Journal, Vol. 81, No. 5, (September 1987).

4. The four regions as defined by the U.S. Census Bureau included the following 50 states and the District of Columbia:
   Northeast: Connecticut, Maine, Massachusetts, New Hampshire, New Jersey, New York, Pennsylvania, Rhose Island, and Vermont.
   Northcentral: Illinois, Indiana, Iowa, Kansas, Michigan, Minnesota, Missouri, Nebraska, North Dakota, Ohio, South Dakota, and Wisconsin.
   South: Alabama, Arkansas, Delaware, District of Columbia, Florida, Georgia, Kentucky, Louisiana, Maryland, Mississippi, North Carolina, Oklahoma, South Carolina, Tennessee, Texas, Virginia, and West Virginia.
   West: Alaska, Arizona, California, Colorado, Hawaii, Idaho, Montana, Nevada, New Mexico, Oregon, Utah, Washingon, and Wyoming.

5. Mark A. Picher, "Fire Fighter Heart Attacks," Fire Command, Vol. 54, No. 7, (July 1987).

6. NFPA Fire Analysis Division, "Estimated Fire Fighters in the U.S., 1985," unpublished, February 1987.